Richard Katalayi Kanyinda
Ziv Kabambi Katambwe
Richard Kamangu Kalonji

OPERATING MODE

Richard Katalayi Kanyinda
Ziv Kabambi Katambwe
Richard Kamangu Kalonji

OPERATING MODE

THE ACN / PCN METHOD

ScienciaScripts

Imprint

Any brand names and product names mentioned in this book are subject to trademark, brand or patent protection and are trademarks or registered trademarks of their respective holders. The use of brand names, product names, common names, trade names, product descriptions etc. even without a particular marking in this work is in no way to be construed to mean that such names may be regarded as unrestricted in respect of trademark and brand protection legislation and could thus be used by anyone.

Cover image: www.ingimage.com

This book is a translation from the original published under ISBN 978-620-6-71259-6.

Publisher:
Sciencia Scripts
is a trademark of
Dodo Books Indian Ocean Ltd. and OmniScriptum S.R.L publishing group

120 High Road, East Finchley, London, N2 9ED, United Kingdom
Str. Armeneasca 28/1, office 1, Chisinau MD-2012, Republic of Moldova, Europe
Printed at: see last page
ISBN: 978-620-7-62223-8

HOW THE ACN / PCN METHOD WORKS

PROCEDURE OF METHOD ACN / PCN

Ziv Kabambi Katambwe[1] , Richard Katalayi Kanyinda[2] and Richard Kamangu Kalonji [3]

[1]Institut du bâtiment et travaux publics (IBTP), Mbujimayi, DR.Congo

[2]Official University of Mbujimayi (UOM), Mbujimayi, RD.Congo

[3]Institut supérieur de commerce (ISC), Mbujimayi, DR. Congo

ABSTRACT

ABSTRACT

Method ACN/PCN is a standardized international system worked out by the International Civil Aviation Organization (ICAO) which aims at providing information on the resistance of

aeronautical roadways (commercial airports and military) and which allows of this fact of judging admissibility of each plane according to its load and the resistance of the roadways.

The ACN represents the aggressiveness of the planes and the PCN the bearing capacity of the structures of roadways accommodating them:plane is acceptable without restriction on an aeronautical roadway if ACN < PCN.if ACN > PCN;particular study must be undertaken to judge admissibility of the plane.¶

KEYWORDS

KEYWORDS : Aerodromes, ACN, PCN, Bearing pressure, ICAO...

SUMMARY

SUMMARY

The ACN / PCN method is an international standardized system developed by the International Civil Aviation Organization (ICAO) to provide information on the strength of
pavements (civil and military airfields), enabling us to assess the suitability of each aircraft in terms of load and pavement resistance.

The ACN represents the aggressiveness of the aircraft and the PCN the bearing capacity of the pavement structures: an aircraft is admissible without restriction on an aeronautical pavement if ACN < PCN, and if ACN >PCN; a special study must be carried out to judge the admissibility of the aircraft.

KEYWORDS: Aerodromes, ACN, PCN, Lift, ICAO...

DEFINITION OF THE ACN / PCN METHOD

1. DEFINITION OF THE ACN / PCN METHOD

By its very nature, the thickness and strength of an aeronautical pavement's subgrade acts as a structural support for the runway, spreading aircraft loads over a larger surface area and thus reducing stresses on the underlying soil.

It helps prevent soil subsidence and compaction, which could compromise runway stability.

The ACN/PCN method was defined by ICAO in 1981, and is binding on all its member states. The method consists in establishing the admissibility of aircraft (characterized by their ACN), according to the bearing capacity of the pavements (characterized by their PCN) of the aerodrome, essentially in the case of flexible pavements.

The ACN (Aircraft Classification Number) **is** a manufacturer-estimated number that expresses the aircraft's aggressiveness on the pavement. This number varies according to the quality of the pavement and is calculated according to the standardized procedures of the ICAO Aerodrome Design Manual.

The PCN (Pavement Classification Number) **is** a number evaluated by the manager which expresses the bearing capacity of the pavement, whether flexible or rigid, for use without

restriction. ICAO does not prescribe a method for calculating the PCN, but leaves it up to States and aerodrome operators. For the sake of comparison, the PCN should be defined according to the same

criteria than ACN. To the best of our knowledge, neither the Swiss civil nor military aviation authorities recommend any particular method at present.

An aircraft is eligible for unrestricted use if its ACN is less than the PCN of the roadway. Occasional derogations are provided for when the ACN exceeds the PCN.

The PORTANCE of a pavement is its ability to support aircraft loads while guaranteeing the integrity of its structure throughout its service life.

CALCULATION OF ACN

2. CALCULATION OF ACN

Standard procedures for calculating ACNs are set out in the Aerodrome Design Manual. They use empirical methods based on experience (numerous tests on experimental runways in the USA), supplemented by theoretical considerations.

On flexible pavements, the CBR (California Bearing Ratio) method is used, based on the punching of the supporting soil and the transmission of loads in a semi-infinite space according to Boussinesq.

For rigid pavements, Westergaard's equations are used for a concrete slab on a Winkler foundation. ACN is expressed as 2 times the permissible load in tonnes on an equivalent single wheel (RSE) inflated to 1.25 MPa, applied 10,000 times.

Determining the ACN of an aircraft involves calculating this equivalent single wheel producing the same effects as the main landing gear of the aircraft in question, as detailed in figure 1.

The ACN of an aircraft varies according to the type of structure (flexible or rigid) and the category of support. ACN is also dependent on tire pressure. However, ACNs are generally supplied without pressure limitation.

Subsequently, ACN is also a linear function of aircraft weight Pt, according to the following formula:

$$ACN = ACN_{min} + (ACN_{max} - ACN_{min}) \cdot \frac{P_t - m}{M - m}$$

With:

Pt: Actual aircraft weight

M: Aircraft weight at maximum load

m: aircraft weight at minimum load

ACN_{min} : ACN has the minimum aircraft load

ACN_{max} : ACN has the maximum load of the aircraft

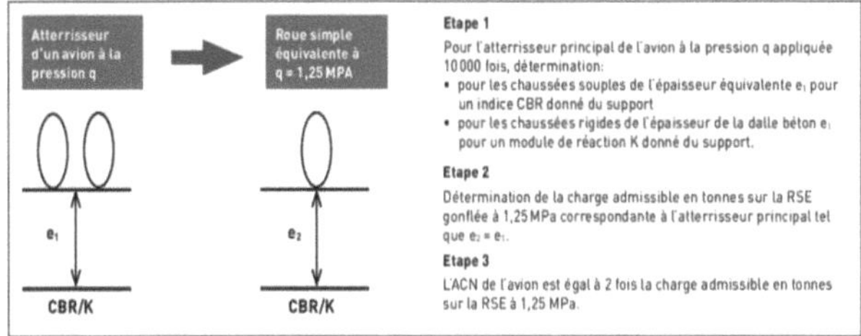

Figure 1: Aircraft ACN calculation process.

Example of ACN determination

A320-200 JUM	Classes de sol							
	Chaussées souples				Chaussées rigides			
Masse de calcul (kg)	A	B	C	D	A	B	C	D
M = 77 400	41	42	47	53	46	49	51	53
m = 40 529	20	20	21	24	22	23	24	25

Figure 2: Example of ACN values for the A320-200 JUM (Source STAC - Winficav database) without tire pressure limitation, with ACN $indices_{min}$ and ACN_{max} for the different categories of support.

DETERMINATION OF NPC

3. DETERMINATION OF NPC

ICAO requires aerodrome managers to declare the load-bearing capacity of aeronautical pavements in terms of PCN indices, without however submitting a method for their determination.

The PCN index is a unitless number with a 4-letter code providing the following information:

PCN code determination

Type de chaussée	Catégorie du support			Pression pneumatique	Méthode d'évaluation
	Code	CBR (chaussée souple)	K (chaussée rigide)		
[-]	[-]	[%]	[MN/m²]	[MPa]	[-]
R = rigide	A: élevé	> 13	> 120	W: pas de limitation	U: Expérience
	B: moyenne	8 .. 13	60 ... 120	X: ‹1.5	
F = flexible	C: bas	4..8	25 60	Y: ‹ 1.0	T: Technique
	D: très bas	‹ 4	‹ 25	Z: ‹ 0.5	

If the load-bearing capacity of a flexible pavement resting on a medium-grade base has been determined at 62 by a technical assessment and without tire pressure limitation,

then the information reported must be: PCN 62/F/B/W/T. There are two evaluation methods for defining the PCN: one is a so-called experience evaluation, the other is a technical evaluation using various procedures detailed below. One is provisional, the other definitive.

DETERMINATION OF NPC BY EXPERIMENT :
U CODE

4.DETERMINATION OF NPC BY EXPERIMENT: U CODE

This method of determining PCN is called "by experience", since it takes into account the influence of past traffic on the roadway: PCN is declared as a function of ACN

of the most aggressive aircraft (called "critical aircraft" or "sizing aircraft") regularly using the aerodrome, subject to acceptable behavior and level of service. The ACN value for this aircraft should be used as the PCN value for the pavement.

The use of this very simple method can only be tolerated for aerodromes with very low traffic and in the case where the predicted frequency of the critical aircraft is similar to the frequency.

known at the time of evaluation.

As this method does not take into account the actual bearing capacity of the pavement, the danger is that it may lead to an overestimation of the PCN index, and thus to the acceptance of traffic for which the PCN index is too low.

that the roadway is undersized.

Increased vigilance of pavement characteristics (cracking, deformation) must be observed in order to detect any increase in deterioration, which would then force us to revise the PCN downwards or opt for the determination of a "Technical" PCN. In any case, use of the U code should be limited in time (2 to 3 years maximum).

DETERMINATION OF TECHNICAL NPC :

5.DETERMINATION OF TECHNICAL CPN: T CODE

A technical evaluation (T-code) of PCN indices first requires the collection of input data:

	Données d'entrée requises	Comment les obtenir?
1	Connaître les caractéristiques des couches de la structure en place (charges admissibles, épaisseurs)	Essai de portance et connaissance des matériaux
2	Définition des charges auxquelles la structure sera soumise	Définition du trafic Définition de la dispersion latérale

The absence of any of the input data (bearing capacity test, knowledge of materials and traffic) makes it impossible to calculate the PCN index reliably.

Some methods are based solely on deflection measurements, giving at best an indication of the pavement's load-bearing capacity. Although relationships have been established between pavement behavior and measured deflection, they remain very approximate. Many external factors, such as knowledge of the pavement structure in place, are not taken into account.

Figure.3 HWD lift meter.

PCN index calculation methods have been classified into three categories: 1. empirical, 2. mechanistic-empirical and 3. Extended" mechanics-empirics. All these methods are based mainly or partly on in situ testing. As a reminder, ACN is also determined empirically.

The second method introduces an analytical approach to material characterization, while remaining close to empirical methods. The third method aims to simulate real-life conditions as closely as possible, both in terms of pavement modelling and the nature of the stresses.

An application example for each of these methods is detailed in the rest of this article. For each of these, the recommended unrestricted traffic consideration is described, but is not included.

does not constitute a generality for the method to which the example relates.

The first methods developed for calculating PCN were empirical ones, namely the CBR and Westergaard methods, also used for calculating ACN.

In the case of flexible pavements, the evaluation should determine both the bearing capacity of the subgrade in terms of CBR and the equivalent thickness of the structure.

For rigid pavements, it involves determining the reaction modulus K and the thickness of the concrete slab, combined with its strength. The French approach is an example of an empirical method.

For flexible pavements, this involves determining the equivalent thickness by assigning theoretical equivalence coefficients to each pavement layer, based on material experiments.

Matériaux	Coefficient d'équivalence
Béton bitumineux à module élevé	2,5
Béton bitumineux aéronautique standard	2
Enrobé à module élevé	1,9
Grave bitume standard	1,5
Grave émulsion	1,2
Grave concassée bien graduée	1
Grave roulée	0,75
Sable	0,5

The equivalent thickness of the pavement is equal to the sum of the thicknesses of each layer, weighted by the equivalence coefficients. These equivalence coefficients are applicable as follows

or are adjusted using the results of deflection measurements, by inverse dimensioning.

Once this data has been collected, it provides 2 options: a flat-rate calculation and an optimized NCP calculation.

The standard calculation consists of calculating the admissible load on a single isolated wheel (RSI [t]) inflated to 0.6 MPa and applied 10,000 times. Conversion to PCN is carried out by multiplying this RSI in tonnes by a coefficient based on the CBR value of the supporting soil.

The optimized calculation takes into account the effects of each aircraft type on the pavement structure. It consists in determining the admissible load applied 10,000 times for each aircraft in the forecast traffic, converted into ACN. The PCN is equal to the sum of the ACNs of the aircraft at their admissible load, weighted by their share of the traffic.

Formula:

$$PCN = \frac{ACN_1 \cdot t_1 + ACN_2 \cdot t_2 + ACN_i \cdot t_i}{\sum_1 t}$$

Avec:
t: nombre équivalent
de mouvements

Optimized calculation is recommended in France, as soon as traffic forecasts are known.

The French approach to determining PCN has the advantage of being in line with ACN's empirical calculation procedures, and remains simple to apply.

However, it has its own limitations:

Theoretical equivalence coefficients do not take into account the actual state of the materials. The reliability of the first input data is therefore insufficient.

Soil deformation is the only criterion considered to define the structure's end-of-life:

- the accuracy of CBR determination can influence the calculated PCN value.

- deformations and stresses in other layers are not taken into account.

It should be noted, however, that this single limit criterion is in line with the standardized procedures for calculating ACN. New landing gear configurations have forced ICAO to adapt ACN calculations by introducing alpha coefficients.

Aircraft lateral dispersion is fixed and invariable. An "empirical" part of the French approach, namely the characterization of materials, has been substituted in the software.

ELMOD (Evaluation of Layer Moduli and Overlay Design) using an analytical approach.

After entering the thicknesses into the software, the elastic moduli of each layer are determined by inverse calculation of the deflections recorded by the HWD (Heavy Weight Deflectometer), based on the Odemark-Boussinesq equivalent thickness principle (MET: Method of Equivalent Thicknesses).

Knowing the permissible stress at ground level and the number of permissible loads, the permissible load on a single wheel equivalent to 1.25 MPa is calculated iteratively.

The PCN index corresponds to 2 times this permissible load in tons.

Finally, the elastic modulus of the soil is converted by correlation into a CBR value in order to be able to publish the category of the substrate (A, B, C or D, cfr. 4. Determination of PCN). We find

the CBR inaccuracy mentioned above.

CALCULATION OF ACN

THE SOFTWARE APPROACH

6. THE SOFTWARE APPROACH

The ELMOD 5 software approach has the advantage of introducing real in-situ material characteristics (first input data), while respecting the equivalent single wheel principle specific to ACN.

However, in accordance with the standard ACN calculation procedures, it only takes into account the deformation of the support, excluding the permissible stresses and deformations in the other layers of the structure. The number of unrestricted traffic passages in ELMOD 5 is entered by the software user, and is therefore not restricted to the 10,000 passages used in the ACN calculation.

As mentioned above, empirical methods are showing their limitations when it comes to adapting to the evolution of materials and new types of loading. To this end, new methods have been developed to provide a more realistic assessment of pavement behavior under repeated loading. On the one hand, traffic modelling has become more rational, since the geometry of the landing gear of forecast traffic, as well as the lateral dispersion of aircraft, is taken into account to calculate the cumulative damage at all points of the pavement's transverse profile.

On the other hand, Burmister or MET modeling is used to calculate stresses and strains in each layer of the pavement structure, and thus define permissible limit criteria for each layer:

Fatigue cracking limit criteria, corresponding to maximum tensile deformations at the base of treated materials, and rutting limit criteria, corresponding to permissible vertical deformations on untreated layers (foundation, soil).

These include the PAVERS and PCASE calculation software packages.

- Layer Elastic Analysis" module, offer the calculation of PCN indices using this method, each of these software packages having its own specificities in the choice of admissible limit criteria. Version ELMOD 6 can also include traffic modeling, but is initially parameterized with the sole limit criterion of ground deformation.

These programs use the number of movements forecast for each aircraft type as unrestricted traffic, reducing it - via equivalences - to a number of passages of the critical aircraft (the aircraft with the highest ACN). The PCN is then expressed as the ACN of the critical aircraft at its permissible weight, for the given number of passages.

The advantage of this software approach is that it determines the bearing capacity, expressed in NCP, based on the characteristics of all the layers of the structure in place. Nevertheless, the admissible limit criteria extended to all layers mean that the method departs from the single criterion on support deformation specific to ACN.

Type de méthode	Empirique		Analytique-empirique	Analytique-empirique «étendue»
Exemple d'application	Approche française «forfaitaire»	Approche française «optimisée»	Logiciel Elmod (Danemark) (Pays-Bas)	Logiciel PAVERS
Qualification des matériaux	Coefficient d'équivalence		Modules élastiques	
Préconisation pour la qualification du sol	Essais CBR / Essais de plaque		Module élastique par calcul inverse	
Critère de rupture	Déformation du sol support			Critère sur les couches couche critique
Détermination du PCN	RSI multiplié par un coefficient fonction du sol support	Somme pondérée des ACN des avions à leur charge admissible	RSE*2	ACN de l'avion critique à sa charge admissible

Figure 4: Comparison of methods used to characterize materials (flexible pavements).

6.1 CHOICE OF NPC DECLARATION

The choice of NCP declaration is the responsibility of the airfield manager, who can introduce risk factors in line with his or her infrastructure management policy and business objectives (traffic development, welcoming new aircraft, etc.).

The PCN is defined as the ACN of the aircraft that can use the roadway for the number of passages expected in unrestricted traffic. Thus, the choice of traffic has an important influence on the calculation of the PCN index: the higher the number of expected passages in unrestricted traffic, the lower the PCN index, and vice versa.

For example, several traffic hypotheses can be studied, so that the operator can choose whether to opt for risky management (low traffic hypothesis $\rightarrow$ high PCN $\rightarrow$ high risk of damage occurring) or cautious management (high traffic hypothesis $\rightarrow$ low PCN $\rightarrow$ low risk of damage occurring).

In addition, since a PCN value can be calculated for each bearing capacity measurement point, a statistical analysis can be carried out to assign a risk rate to the declared PCN value, corresponding to the percentage of the pavement surface undersized for the forecast traffic. A risk rate of 20% is usually attributed to normal management, while a rate of 50% corresponds to management at risk, i.e. the probability that half of the pavement surface will experience significant (structural) deterioration before reaching the end of its service life.

6.2 PROSPECTS FOR THE ACN/PCN METHOD

Although empirical methods are still widely used, mechanical-empirical methods are making progress, thanks to their ability to incorporate variations in traffic criteria and material types into the calculation, leading to a more rational assessment of permissible loads and therefore pavement PCN. ICAO member states are aware of this, and work is underway to switch from the ACN/PCN system to analytical-empirical methods.

The working group was set up to propose modifications to the ACN calculation, taking into account the responses of elastic linear multi-layer structures to stresses.

This mechanical approach would have the particular advantage of considering the effects of multi-wheel landing gear. However, ground deformation would remain the sole failure criterion for defining the admissible load.

This would mean abandoning the CBR Westergaard methods and re-evaluating aircraft ACNs according to the new standardized calculation. At the same time, a new

of the PCN calculation, closer to the mechanic-empirical methods described, could be imposed by ICAO to take account of these modifications to the ACN calculation.

These changes should then be seen as a necessary improvement for more efficient management of aeronautical pavements.

Pending this standardization, which is desirable in order to facilitate the understanding of managers who have to publish their PCN indices, Infralab SA has opted for PCN calculation using ELMOD software, i.e. for a mechanical-empirical method that is transitional in the current context.

This method complies as closely as possible with the ACN/PCN system described by ICAO, while incorporating mechanical criteria for material characterization.

CONCLUSION

7. CONCLUSION

The ACN/PCN method is a tool for managing a platform's aeronautical pavements. This can be achieved by precise counting of aircraft movements, both in terms of actual weight and taxi path. It also enables us to know whether an aircraft's access authorization to an aerodrome can be accepted or whether it should be refused.

The purpose of this tool is to provide managers with a means of monitoring the "load-bearing potential" of their pavements and planning future investments.

BIBLIOGRAPHICAL REFERENCES

8. BIBLIOGRAPHICAL REFERENCES

1. STBA. (1988). Practical guide to using the ACN/PCN method.

2. ICAO, Annex 14 - Aerodromes, Volume I - Aerodrome design and operation
Aerodromes, 5th edition, July 2009.

3. ICAO, Aerodrome Design Manual, Part 3, Pavements, 2nd edition - 1983.

4. Service Technique de l'Aviation Civile - ITAC - Chap. 5, aeronautical pavement design.

5. STAC - Lift assessment http://www.stac.aviation-civile.gouv.fr/chaussee/portance.php

6. Service Technique de l'Aviation Civile - ITAC - Chap. 8, Aeronautical pavement management - ACN/PCN method

7. Airbus Engineering - Airport pavements International Conference in High Tatras, Slovak Republic (21/05/2012) - ICAO-UPDATE, ACN/PCN

8. Modification des ACN de certains aéronefs suite à la décision de l'OACI du16 octobre 2007, STAC, juin 2008, http://www.stac.aviation-civile.gouv.fr/publications/gnt- chaus.php

9. Mouri L. and Bessadat H., 1994. Etude de dimensionnement des chaussées aéronautiques type

sud. PFE, final year project, Université des Sciences et de la Technologie d'Oran.

10. Uge P., Gravoise A. and Lemaire J.N., June 1976. Le comportement en fatigue des enrobés bitumineux : influence du liant, Revue Générale des Routes et des Aérodromes n° 521.

11. Dynatest International A/S, ELMOD6 software guide "Evaluation of Layer Moduli
and Overlay Design".

CALCULATION OF ACN

TABLE OF CONTENTS

yes

I want morebooks!

Buy your books fast and straightforward online - at one of world's fastest growing online book stores! Environmentally sound due to Print-on-Demand technologies.

Buy your books online at
www.morebooks.shop

Kaufen Sie Ihre Bücher schnell und unkompliziert online – auf einer der am schnellsten wachsenden Buchhandelsplattformen weltweit! Dank Print-On-Demand umwelt- und ressourcenschonend produzi ert.

Bücher schneller online kaufen
www.morebooks.shop